A Review of the DOE Plan for
U.S. Fusion Community Participation in the ITER Program

Committee to Review the U.S. ITER Science Participation Planning Process

Plasma Science Committee
Board on Physics and Astronomy
Division on Engineering and Physical Sciences

NATIONAL RESEARCH COUNCIL
OF THE NATIONAL ACADEMIES

THE NATIONAL ACADEMIES PRESS
Washington, D.C.
www.nap.edu

THE NATIONAL ACADEMIES PRESS   500 Fifth Street, N.W.   Washington, DC 20001

NOTICE: The project that is the subject of this report was approved by the Governing Board of the National Research Council, whose members are drawn from the councils of the National Academy of Sciences, the National Academy of Engineering, and the Institute of Medicine. The members of the committee responsible for the report were chosen for their special competences and with regard for appropriate balance.

This study was supported by Grant No. DE-FG02-07ER54924 between the National Academy of Sciences and the Department of Energy. Any opinions, findings, conclusions, or recommendations expressed in this publication are those of the author(s) and do not necessarily reflect the views of the organizations or agencies that provided support for the project.

This report was prepared as an account of work sponsored by an agency of the United States Government. Neither the United States Government nor any agency thereof, nor any of their employees, makes any warranty, express or implied, or assumes any legal liability or responsibility for the accuracy, completeness, or usefulness of any information, apparatus, product, or process disclosed, or represents that its use would not infringe privately owned rights. Reference herein to any specific commercial product, process, or service by trade name, trademark, manufacturer, or otherwise does not necessarily constitute or imply its endorsement, recommendation, or favoring by the United States Government or any agency thereof. The views and opinions of authors expressed herein do not necessarily state or reflect those of the United States Government or any agency thereof.

Cover: Computer-generated image of the ITER magnet system, including the plasma-facing internal components. For scale, note the size of the person represented at bottom-center. Image available at http://www.iter.org/ and reprinted with permission of the ITER Organization. Cover design by Steven Coleman.

International Standard Book Number-13:   978-0-309-12475-1
International Standard Book Number-10:   0-309-12475-1

Copies of this report are available from the Board on Physics and Astronomy, National Research Council, 500 Fifth Street, N.W., Washington, DC 20001; Internet, http://www.national-academies.org/bpa. Additional copies of this report are available from the National Academies Press, 500 Fifth Street, N.W., Lockbox 285, Washington, DC 20055 or (800) 624-6242 or (202) 334-3313 (in the Washington metropolitan area); Internet, http://www.nap.edu.

Copyright 2009 by the National Academy of Sciences. All rights reserved.

Printed in the United States of America

THE NATIONAL ACADEMIES
Advisers to the Nation on Science, Engineering, and Medicine

The **National Academy of Sciences** is a private, nonprofit, self-perpetuating society of distinguished scholars engaged in scientific and engineering research, dedicated to the furtherance of science and technology and to their use for the general welfare. Upon the authority of the charter granted to it by the Congress in 1863, the Academy has a mandate that requires it to advise the federal government on scientific and technical matters. Dr. Ralph J. Cicerone is president of the National Academy of Sciences.

The **National Academy of Engineering** was established in 1964, under the charter of the National Academy of Sciences, as a parallel organization of outstanding engineers. It is autonomous in its administration and in the selection of its members, sharing with the National Academy of Sciences the responsibility for advising the federal government. The National Academy of Engineering also sponsors engineering programs aimed at meeting national needs, encourages education and research, and recognizes the superior achievements of engineers. Dr. Charles M. Vest is president of the National Academy of Engineering.

The **Institute of Medicine** was established in 1970 by the National Academy of Sciences to secure the services of eminent members of appropriate professions in the examination of policy matters pertaining to the health of the public. The Institute acts under the responsibility given to the National Academy of Sciences by its congressional charter to be an adviser to the federal government and, upon its own initiative, to identify issues of medical care, research, and education. Dr. Harvey V. Fineberg is president of the Institute of Medicine.

The **National Research Council** was organized by the National Academy of Sciences in 1916 to associate the broad community of science and technology with the Academy's purposes of furthering knowledge and advising the federal government. Functioning in accordance with general policies determined by the Academy, the Council has become the principal operating agency of both the National Academy of Sciences and the National Academy of Engineering in providing services to the government, the public, and the scientific and engineering communities. The Council is administered jointly by both Academies and the Institute of Medicine. Dr. Ralph J. Cicerone and Dr. Charles M. Vest are chair and vice chair, respectively, of the National Research Council.

www.national-academies.org

COMMITTEE TO REVIEW THE U.S. ITER SCIENCE PARTICIPATION PLANNING PROCESS

PATRICK L. COLESTOCK, Los Alamos National Laboratory, *Chair*
ROGER D. BENGTSON, University of Texas at Austin
JAMES E. BRAU, University of Oregon
CARY B. FOREST, University of Wisconsin
STEPHEN HOLMES, Fermi National Accelerator Laboratory
GEORGE J. MORALES, University of California at Los Angeles
THOMAS M. O'NEIL, University of California at San Diego
TONY S. TAYLOR, General Atomics
DENNIS G. WHYTE, Massachusetts Institute of Technology
MICHAEL C. ZARNSTORFF, Princeton University

Staff

DONALD C. SHAPERO, Director, Board on Physics and Astronomy
TIMOTHY I. MEYER, Senior Program Officer (August 2006–September 2007)
DAVID LANG, Program Officer (from October 2007)
MERCEDES ILAGAN, Administrative Assistant (October 2007–February 2008)
CARYN KNUTSEN, Program Associate (from March 2008)
BETH C. DOLAN, Financial Associate

PLASMA SCIENCE COMMITTEE

RICCARDO BETTI, University of Rochester, *Chair*
MICHAEL R. BROWN, Swarthmore College
LINDA M. CECCHI, Sandia National Laboratories
PATRICK L. COLESTOCK, Los Alamos National Laboratory
S. GAIL GLENDINNING, Lawrence Livermore National Laboratory
VALERY GODYAK, OSRAM Sylvania, Inc.
IAN H. HUTCHINSON, Massachusetts Institute of Technology
CHADRASHEKHAR JOSHI, University of California at Los Angeles
ELIOT QUATAERT, University of California at Berkeley
EDWARD THOMAS, JR., Auburn University
MICHAEL C. ZARNSTORFF, Princeton University
THOMAS H. ZURBUCHEN, University of Michigan

Staff
DONALD C. SHAPERO, Director, Board on Physics and Astronomy
DAVID B. LANG, Program Officer
CARYN J. KNUTSEN, Program Associate

BOARD ON PHYSICS AND ASTRONOMY

MARC A. KASTNER, Massachusetts Institute of Technology, *Chair*
ADAM S. BURROWS, University of Arizona, *Vice Chair*
JOANNA AIZENBERG, Harvard University
JAMES E. BRAU, University of Oregon
PHILIP H. BUCKSBAUM, Stanford University
PATRICK L. COLESTOCK, Los Alamos National Laboratory
RONALD C. DAVIDSON, Princeton University
ANDREA M. GHEZ, University of California at Los Angeles
PETER F. GREEN, University of Michigan
LAURA H. GREENE, University of Illinois at Urbana-Champaign
MARTHA P. HAYNES, Cornell University
JOSEPH HEZIR, EOP Group, Inc.
MARK B. KETCHEN, IBM T.J. Watson Research Center
ALLAN H. MacDONALD, University of Texas at Austin
PIERRE MEYSTRE, University of Arizona
HOMER A. NEAL, University of Michigan
JOSE N. ONUCHIC, University of California at San Diego
LISA J. RANDALL, Harvard University
CHARLES V. SHANK, Janelia Farm, Howard Hughes Medical Institute
THOMAS N. THEIS, IBM T.J. Watson Research Center
MICHAEL S. TURNER, University of Chicago
MICHAEL C.F. WIESCHER, University of Notre Dame

Staff

DONALD C. SHAPERO, Director
MICHAEL H. MOLONEY, Associate Director
ROBERT L. RIEMER, Senior Program Officer
JAMES LANCASTER, Program Officer
DAVID B. LANG, Program Officer
CARYN J. KNUTSEN, Program Associate
ALLISON M. McFALL, Senior Program Assistant
BETH C. DOLAN, Financial Associate

Preface

The development of a plan for the participation of the U.S. fusion community in the ITER program was mandated by the Energy Policy Act of 2005 (EPAct; Public Law 109-58, August 8, 2005). The EPAct, in Section 972 (c)(4)(B), also directed that, after completion of the plan, the U.S. Department of Energy (DOE) request an external review of its content. Accordingly, on August 10, 2006, the DOE under secretary for science submitted the completed plan to the National Academy of Sciences for review (see Appendix A). In response, the National Research Council (NRC) organized a committee to review the DOE plan with the following charge:

> The committee will prepare a short report addressing the following tasks:
>
> 1. Review the document "Planning for U.S. Fusion Community Participation in the ITER Program." Determine whether the plan provides a good initial outline for effective participation of U.S. plasma scientists in research at ITER.
>
> 2. Evaluate the following required elements of the plan: (1) an agenda for U.S. research at ITER, (2) methodologies to evaluate ITER's contribution to progress toward a power source, (3) description of the anticipated relationship between the U.S. ITER research program and the overall U.S. fusion program.
>
> 3. The committee will recommend next steps in the development of the plan, including: (a) appropriate elements and/or goals for the

plan; (b) procedures to facilitate further development of the plan; and (c) metrics for measuring progress in establishing robust U.S. participation in the ITER research program.

The Committee to Review the U.S. ITER Science Participation Planning Process was appointed on October 1, 2007, and met in Washington, D.C., on December 14–15, 2007. Soon after, the FY2008 Consolidated Appropriations Act (Public Law 110-161, December 26, 2007) became law, under which U.S. contributions for ITER were unexpectedly eliminated. Although this committee was not specifically asked to assess the implications of the FY2008 budget, it believes that the budget will necessarily affect U.S. researchers' ability to participate fully in the ITER program, and it therefore felt obliged to address this issue.

This report reviews and evaluates the DOE plan and the status of DOE planning based on the above criteria, and recommends next steps in the development of the plan. The committee observes that domestic planning activities have been effective thus far. However, as the ITER project progresses, the organizational landscape will likely change, as will the developing international research agenda. The committee therefore presents a snapshot of the ITER program as it exists at the present time. The full value of the committee's guidance lies in its recommended elements and procedures to help position the United States to maximize its participation in and reward from the important international scientific and technical endeavor embodied in ITER.

The committee thanks the guest speakers at its December 14, 2007, meeting, including Kathryn Beers, Office of Science and Technology Policy; Earl Marmar, Massachusetts Institute of Technology; Stanley Milora, Oak Ridge National Laboratory; Erol Oktay, Department of Energy; Ned Sauthoff, Oak Ridge National Laboratory; and James Van Dam, University of Texas at Austin and U.S. Burning Plasma Organization. Special thanks are due to our foreign colleagues who participated in the meeting despite the long distances, namely, David Campbell, ITER Organization; Shinzaburu Matsuda, Japan Atomic Energy Agency; and Jerome Pamela, European Fusion Development Agreement. The committee greatly appreciates the time and effort that all of these individuals put into preparing their remarks and participating in discussions.

Patrick L. Colestock, *Chair*
Committee to Review the U.S. ITER
Science Participation Planning Process

Acknowledgment of Reviewers

This report has been reviewed in draft form by individuals chosen for their diverse perspectives and technical expertise, in accordance with procedures approved by the National Research Council's Report Review Committee. The purpose of this independent review is to provide candid and critical comments that will assist the institution in making its published report as sound as possible and to ensure that the report meets institutional standards for objectivity, evidence, and responsiveness to the study charge. The review comments and draft manuscript remain confidential to protect the integrity of the deliberative process. We wish to thank the following individuals for their review of this report:

Gordon Baym, University of Illinois at Urbana-Champaign,
Michael R. Brown, Swarthmore College,
Steven C. Cowley, University of California at Los Angeles,
Ronald C. Davidson, Princeton University,
Joseph Hezir, EOP Group, Inc.,
Charles F. Kennel, University of California at San Diego,
Christopher Llewellyn-Smith, United Kingdom Atomic Energy Authority–Culham Division,
David Meyerhofer, University of Rochester,
John Peoples, Jr., Fermi National Accelerator Laboratory, and
Clifford Surko, University of California at Berkeley.

Although the reviewers listed above have provided many constructive comments and suggestions, they were not asked to endorse the conclusions or recommendations, nor did they see the final draft of the report before its release. The review of this report was overseen by John F. Ahearne of Sigma Xi and Duke University. Appointed by the National Research Council, he was responsible for making certain that an independent examination of this report was carried out in accordance with institutional procedures and that all review comments were carefully considered. Responsibility for the final content of this report rests entirely with the authoring committee and the institution.

Contents

EXECUTIVE SUMMARY   1

1   INTRODUCTION   5
History of the ITER Project, 5
The Present ITER Project, 7
Recent U.S. Developments, 8
Origin of This Study, 10

2   EVALUATION OF THE CURRENT DOE PLAN FOR U.S.   11
PLASMA SCIENCE COMMUNITY PARTICIPATION IN ITER
Assessment of Organization and Planning of the U.S. ITER
    Effort, 11
  Key Structural Elements of U.S. Participation in ITER, 13
  Comparison to Analogous Efforts of Other ITER Partners, 15
  Assessment of the U.S. Research Agenda at ITER, 16
  Alignment with DOE/OFES Goals and Previous NRC and
    FESAC Advice, 19
  Areas of Concern, 20
Assessment of Methodologies to Evaluate ITER's Contribution
    to Progress Toward a Power Source, 23
Relationship of the U.S. Fusion Program to the U.S. ITER
    Research Program, 24

3 RECOMMENDED ELEMENTS FOR FUTURE DEVELOPMENT 26
   OF THE DOE PLAN FOR U.S. PLASMA SCIENCE
   COMMUNITY PARTICIPATION IN ITER
   Recommended Goals of U.S. ITER Planning Activities, 26
   Recommended Procedures to Facilitate Further Development
       of the DOE Plan, 27
   Recommended Metrics for Measuring Robust U.S. Participation
       in the ITER Research Program, 29
     Metrics Included in the DOE Plan, 29
     Recommended Additional Metrics, 31

APPENDIXES

A Letter of Request from the U.S. Department of Energy 37
B Meeting Agenda 39

Executive Summary

ITER, a planned next-generation fusion research facility, presents the United States and its international partners with the opportunity to explore new and exciting frontiers of plasma science while bringing the promise of fusion energy closer to reality. The ITER project has garnered the commitment and will draw on the scientific potential of seven international partners—China, the European Union, India, Japan, the Republic of Korea, Russia, and the United States—countries that represent more than half of the world's population. The success of ITER will depend on each partner's ability to fully engage in the scientific and technological challenges posed by advancing the understanding of fusion.

The National Research Council's Committee to Review the U.S. ITER Science Participation Planning Process was asked to assess the current U.S. Department of Energy (DOE) plan for U.S. fusion community participation in ITER, evaluate the plan's elements, and recommend appropriate goals, procedures, and metrics for consideration in the future development of the plan.[1] The committee found that:

- The 2006 DOE plan for U.S. participation in ITER is operating and has proven effective in beginning to coordinate U.S. research activities and the development of the ITER program. U.S. scientists have

---

[1] U.S. Burning Plasma Organization, *Planning for U.S. Fusion Community Participation in the ITER Program*, June 7, 2006. The DOE plan is available at http://www.ofes.fusion.doe.gov/News/EPAct_final_June06.pdf, last viewed on July 22, 2008.

been well engaged in the planning for ITER, and the United States should endeavor to maintain this level of activity. The plan, in its current form, is well aligned with DOE Office of Fusion Energy Sciences goals.
- The U.S. ITER research program is at least as organizationally and technically mature as that of the other ITER participants at the time of this writing.[2]
- The U.S. research program for ITER as described in the DOE plan is appropriate and justified, and the committee notes that the domestic program will evolve as the international research program is developed. U.S. involvement in developing the research program for ITER will be crucial to the realization of U.S. fusion research goals.
- The committee underscores as its greatest concern the uncertain U.S. commitment to ITER at the present time. Fluctuations in the U.S. commitment to ITER will undoubtedly have a large negative impact on the ability of the U.S. fusion community to influence the developing ITER research program, to capitalize on research at ITER to help achieve U.S. fusion energy goals, to participate in obtaining important scientific results on burning plasmas from ITER, and to be an effective participant in and beneficiary of future international scientific collaborations.
- Consistent with previous National Research Council and Fusion Energy Sciences Advisory Committee reports, the committee emphasizes that a vigorous and strategically balanced domestic program is required to ensure that U.S. participation in ITER is successful and valuable for the U.S. fusion program.
- The DOE plan for U.S. participation in ITER includes well-thought-out metrics for measuring progress toward development of fusion energy as a power source.
- The DOE plan includes well-thought-out metrics to measure the robustness of U.S. participation in the ITER program.

Based on these findings, the committee makes the following recommendations:

- **The Department of Energy should take steps to seek greater U.S. funding stability for the international ITER project to ensure that the United States remains able to influence the developing ITER research program, to capitalize on research at ITER to help achieve U.S. fusion energy goals, to participate in obtaining**

---

[2] As of April 8, 2008.

important scientific results on burning plasmas from ITER, and to be an effective participant in and beneficiary of future international scientific collaborations.

- Important considerations that are not reflected in the current DOE plan for U.S. participation in ITER should be addressed during the further development of the DOE plan. These considerations include:
  —Existing gaps in planning for a Demonstration Power Plant,
  —Dissemination of information on and the results of ITER research activities to the broader scientific community, and
  —Planning for the recruitment and training of young scientists and engineers.

- The committee recommends that the following goals be adopted as the foundation of DOE planning activities for U.S. participation in ITER:
  —Ensuring broad academic and industry participation in ITER,
  —Enabling the United States to contribute substantially to and reap the rewards from ITER, and
  —Recruiting and training young fusion scientists and engineers.

- The committee recommends the following procedures to accomplish the U.S. planning goals recommended above, and to facilitate the further development of the DOE plan:
  —DOE should create a long-term strategic plan for the U.S. burning plasma fusion program within the context of global fusion energy development activities.
  —The U.S. Burning Plasma Organization should continue to be an essential point of communication, and serve as a home team to encourage broad cooperation and collaboration among all U.S. participants in the ITER project.
  —DOE should maintain a vibrant domestic fusion program through strong support for basic research and facilities.
  —The DOE plan for U.S. participation in ITER should consider what capabilities exist and need to exist at U.S. plasma science facilities.
  —The DOE plan should consider the needed operating availability of domestic tokamaks.

- The committee recommends that the following five metrics be considered for inclusion during the future development of the DOE plan for U.S. fusion community participation in ITER:

—Periodic evaluation by expert and knowledgeable members of the scientific, engineering, and industrial community regarding the U.S. return on its ITER investment.
—Periodic assessments by independent, external bodies of the effectiveness of domestic project management.
—Balance in the fraction of U.S. published research conducted on ITER according to authors' institutional affiliations (university, national laboratory, and industry).
—Number of research and technology publications documenting results obtained on ITER that are cited by or produced in collaboration with U.S. researchers, students, and technologists across U.S. plasma science and physics.
—Achievement of predictive capability, to be evaluated by peer review.

1

Introduction

HISTORY OF THE ITER PROJECT

The idea to utilize a controlled, sustainable, magnetically confined plasma to generate energy by fusing together light nuclei was first envisioned in the 1950s following research stemming from the Manhattan Project. In 1958, fusion energy research was declassified, triggering a decade of nascent research efforts around the world. In 1968, the Soviet Union reported a major breakthrough in magnetically confined fusion—a concept for a confinement device called a "tokamak," an acronym based on Russian words for toroidal magnetic chamber. Following this breakthrough, fusion research developed rapidly, consistently doubling tokamak performance every year, as countries competed to improve the performance of the tokamak concept over successive generations of experiments.

As technical capabilities expanded, worldwide interest grew regarding the potential benefits of fusion research for society. Harnessing fusion energy for domestic energy production became an element of U.S. energy policy during the energy crisis of the 1970s. As the crisis continued, President Jimmy Carter's administration highlighted the importance of fusion energy in the Magnetic Fusion Energy Engineering Act of 1980, which specified aggressive pursuit of fusion research. But just when the act took effect, the energy crisis began to retreat owing to various world events. As a consequence, the recommendations of the 1980 act were never implemented by the U.S. government. Later, at the Geneva Summit in 1985, the United States joined the Soviet Union, the European Union (EU), and Japan to undertake joint design of a tokamak experimental reactor. This design provided the early foundations for the current ITER project.

By the mid-1990s, two tokamak devices achieved the generation of controlled fusion power of more than 10 megawatts for a period on the order of several seconds. The devices were the Tokamak Fusion Test Reactor (TFTR) in Princeton, New Jersey, and the Joint European Torus (JET) in the United Kingdom. The experimental milestones achieved at these facilities in the confinement, heating, and control of the plasma and the first use of tritium fuel were significant. Scientifically a critical finding was that the energetic helium ions produced by the deuterium-tritium (D-T) fusion reaction were well confined and behaved as expected; that is, they "gave back" essentially all their energy to the plasma itself. These experiments provided the technical and scientific confidence that a burning plasma could be achieved in a next-generation device, the device currently designated as ITER. In such "burning plasma" devices the 20 percent of the energy generated by the fusion reactions found in the He ions mentioned above is used to maintain the necessary high temperatures—that is, the fusion reactions will self-heat and sustain the plasma. This is the fundamental feature of an energy-producing tokamak plasma that will be found in fusion reactors, but not in present devices.

Although the United States was one of the original ITER partners, in 1998 Congress ordered DOE to withdraw from the international collaboration. In spite of the U.S. withdrawal, partners in Europe, Russia, and Japan continued to advance the design of the project. These efforts, although they resulted in a slight descoping of technical objectives, led to the present ITER design that provides access to burning plasma regimes at a reduced cost. In parallel, the U.S. fusion community held a series of workshops that demonstrated broad support for advancing a burning plasma experiment. Several burning plasma options were examined, and the U.S. community gave the new ITER design a favorable technical assessment. The community also noted that the ITER project had adopted changes advocated by the United States. Motivated by the renewed prospect of a positive next step in magnetic fusion research, in 2002 the DOE Fusion Energy Sciences Advisory Committee voiced its support for a renewal of U.S. participation in ITER negotiations. Similarly, the U.S. National Research Council's Burning Plasma Assessment Committee in its 2002 interim letter report reaffirmed this recommendation to rejoin talks and stated in its subsequent full report that "the U.S. fusion program, after many years of research, is poised to take a major step toward its energy goal. It is clear that a burning plasma experiment is a necessary step on the road to fusion energy and of scientific and technical interest to the U.S.

fusion program and beyond."[1] On January 30, 2003, President George W. Bush announced that the United States would rejoin the collaboration.[2] In addition to the original 1996 members—Russia, the United States, the EU, and Japan—the project also included as new members the People's Republic of China and the Republic of Korea (followed by India in 2005), indicating the broad international appeal of and support for the project. In November 2003, Secretary of Energy Spencer Abraham announced that ITER would be the top priority in the 20-year facility development plan of the DOE Office of Science.

Although complicated, the history of U.S. participation in the ITER project highlights the project's resiliency, both in terms of its science appeal and as a groundbreaking international collaboration. Lessons learned from earlier international collaborations, such as the Large Hadron Collider, have contributed to effective organization of the ITER project. In fact, ITER is being considered as a model for future large-scale, international science projects.

THE PRESENT ITER PROJECT

As stated in the ITER Joint Implementation Agreement (JIA), the objective of the ITER project is "to demonstrate the scientific and technological feasibility of fusion energy for peaceful purposes, an essential feature of which would be achieving sustained fusion power generation."[3] According to the ITER Web site, "ITER will accomplish this objective by demonstrating high power amplification and extended burn of deuterium-tritium plasmas, with steady-state as an ultimate goal, by demonstrating technologies essential to a reactor in an integrated system, and by performing integrated testing of the high-heat-flux and nuclear components required to utilize fusion energy for practical purposes."[4]

The current plan is that construction of ITER will begin in 2008. ITER seeks to achieve its first plasma in 2018 and is expected to operate for 20 years. It aims to produce 500 MW of fusion power for 400 seconds by 2024. Commensurate with agreed-on levels of involvement in the ITER

---

[1]National Research Council, *Letter Report: Burning Plasma Assessment (Phase 1)*, The National Academies Press, Washington, D.C., 2002; National Research Council, *Burning Plasma: Bringing a Star to Earth*, The National Academies Press, Washington, D.C., 2004, p. 38.

[2]George W. Bush, *Promoting Energy Independence Through Cooperative Research to Develop Fusion Energy*, Presidential Initiative, released January 30, 2003.

[3]ITER Organization, *Agreement on the Establishment of the ITER International Fusion Energy Organization for the Joint Implementation of the ITER Project*, Article 2, November 21, 2006. Available at http://www.iter.org/JIA_text.htm, last viewed March 6, 2008.

[4]As defined on the ITER Web site at http://www.iter.org/Objectives.htm, last viewed March 6, 2008.

project, the host, the EU, will provide 5/11 (45.4 percent) and the six nonhosts will each provide 1/11 (9.1 percent) of the in-kind contributions for construction, which for the most part consist of components for the machine.

The formal site selection process for ITER began with Canada's proposal to locate the experiment at Clarington, Ontario, in 2001, followed by proposals for a Japanese site at Rokkasho-Mura, a Spanish site at Vandellos, and a French site at Cadarache. The EU decided to consolidate the European site proposals to a single one at Cadarache, which ultimately proved successful on June 28, 2005.

On November 21, 2006, the United States and its international partners signed the International Fusion Energy Agreement, cementing the seven member countries' participation in the project. Less than a year later, on October 24, 2007, with the signatures of the ITER parties, the ITER Organization was officially created, and the United States, along with its six foreign collaborators, became official, fully participating members. The purpose of the ITER Organization is "to provide for and to promote cooperation among the Members . . . on the ITER Project."[5] As it becomes operational, the ITER Organization will coordinate the construction and operation of ITER and will interface with the seven nations involved in the project. In addition, the EU and Japan negotiated a separate bilateral agreement (the "Broader Approach" agreement) to jointly construct and operate a number of fusion facilities in parallel with ITER to be sited in Japan.[6]

RECENT U.S. DEVELOPMENTS

Since the U.S. decision to participate, domestic progress on the project has been smooth until recently. In the Energy Policy Act of 2005 (Public Law 109-58, August 8, 2005), Congress authorized negotiation of "an agreement for United States participation in the ITER," and participation in ITER is identified by the DOE Office of Science as its top priority for the next 20 years.[7]

However, in the FY2008 U.S. Consolidated Appropriations Act (Public Law 110-161, December 26, 2007), funding for the project was nearly eliminated for the year. Although DOE had requested from Congress

---

[5]ITER Organization, *Agreement on the Establishment of the ITER International Fusion Energy Organization for the Joint Implementation of the ITER Project,* Article 2, November 21, 2006. Available at http://www.iter.org/JIA_text.htm, last viewed March 6, 2008.

[6]See "The Broader Approach," available at http://www.iter.org/Broad.htm, last viewed July 22, 2008.

[7]U.S. Department of Energy, *Four Years Later: An Interim Report on Facilities for the Future of Science: A Twenty-Year Outlook,* Washington, D.C., August 2007, p. 8.

INTRODUCTION

"funding of $160.0 million in FY 2008,"[8] the FY2008 budget as appropriated allocates "$0 for the U.S. contribution to ITER, and $10,724,000 for Enabling R&D for ITER," adding that "[f]unding may not be reprogrammed from other activities within Fusion Energy Sciences to restore the U.S. contribution to ITER."[9] This action eliminated funding for the U.S. in-kind equipment contributions to ITER; funding for U.S. personnel to work at the ITER site; cash for the U.S. share of common expenses such as infrastructure, hardware assembly, and installation; and contingency funds for the ITER Organization for FY2008. U.S. financial participation in the international project remains suspended at the time of this report's writing. Although U.S. funding for the project has wavered, DOE Under Secretary for Science Raymond Orbach, in a letter to ITER Organization Director General Kaname Ikeda, stated "that the U.S. is firmly committed to meeting our obligations under the ITER Joint Implementation Agreement (JIA) and that we are doing everything possible to rectify the situation."[10] For FY2008, at least, the implications of the FY2008 appropriations as stated in Dr. Orbach's letter are that "there will be some limitations in our ability to fully participate in ITER activities" but that the United States will remain engaged in key technical, scheduling, and planning activities.

U.S. participation in ITER in FY2008 will be at a minimal level, and its cash and in-kind procurement contributions will be zero. The lack of the anticipated funding has implications for the U.S. ability to participate in and influence the project, given that the U.S. ITER Project Office has been reduced to a core team. It is also worth noting that the promised contributions will remain due under the JIA, as will contributions in the out-years, and that DOE will have to make up the difference.

The President's FY2009 budget request to Congress includes $214.5 million for the ITER project that, if appropriated, will restore U.S. participation in FY2009. However, support for the project in the subsequent out-years is not guaranteed. It will take strong leadership from the U.S. executive and legislative branches to ensure the ITER project's long-term health and success.

---

[8]U.S. Department of Energy, *FY2008 Congressional Budget Request,* Washington, D.C., 2008, p. 72.

[9]*U.S. Consolidated Appropriations Act for FY2008*, Public Law 110-161, Washington, D.C., December 26, 2007.

[10]U.S. Department of Energy, Letter from Under Secretary for Science Raymond Orbach to ITER Organization Director General Kaname Ikeda, January 10, 2008.

ORIGIN OF THIS STUDY

In Section 972 (c)(4)(A) of the Energy Policy Act of 2005, Congress directed that DOE

> in consultation with the Fusion Energy Sciences Advisory Committee, . . . develop a plan for the participation of United States scientists in the ITER that shall include:
>
> (i) the United States research agenda for the ITER;
>
> (ii) methods to evaluate whether the ITER is promoting progress toward making fusion a reliable and affordable source of power; and
>
> (iii) a description of how work at the ITER will relate to other elements of the United States fusion program.

In February 2006, DOE asked the U.S Burning Plasma Organization (USBPO) to develop that plan. The resulting report, *Planning for U.S. Fusion Community Participation in the ITER Program*,[11] completed in June 2006, represents an important first step in organizing the U.S. ITER Project Office and the plasma science community to successfully participate in the project. The plan was submitted to Congress by DOE on August 10, 2006.

In Section 972 (c)(4)(B) of the Energy Policy Act of 2005, DOE was directed to request a review of the plan by the National Academy of Sciences. At DOE's request (see Appendix A), the National Research Council's Committee to Review the U.S. ITER Science Participation Planning Process was thus convened and asked to review and evaluate the current DOE plan, *Planning for U.S. Fusion Community Participation in the ITER Program*, and to recommend elements for future development of the plan for U.S. plasma science participation in the ITER project.

---

[11]U.S. Burning Plasma Organization, *Planning for U.S. Fusion Community Participation in the ITER Program*, June 7, 2006. Available at http://www.ofes.fusion.doe.gov/News/EPAct_final_June06.pdf, last viewed July 22, 2008.

2

Evaluation of the Current DOE Plan for U.S. Plasma Science Community Participation in ITER

ASSESSMENT OF ORGANIZATION AND PLANNING OF THE U.S. ITER EFFORT

The Department of Energy (DOE) plan[1] provides defined structures for organizing the participation of U.S. researchers in ITER research during the construction phase, and a phased U.S. research agenda for ITER. The plan also identifies mechanisms for adapting and advancing the plan as ITER develops (see Figures 2.1 and 2.2). In the period since August 2006 when the plan was submitted to Congress, the structures and mechanisms that it describes have been established and are operating. In addition, the ITER agreement came into force, the international ITER Organization was established, and an international technical review of the ITER design was conducted.

The DOE plan provides effective mechanisms and guidance for supporting U.S. participation in ITER research, addressing the U.S. research agenda. The plan has been elaborated and built upon in subsequent planning processes, including the ongoing Fusion Energy Sciences Advisory Committee (FESAC) strategic planning and U.S. participation in ITER Organization (IO) research planning.

---

[1] U.S. Burning Plasma Organization, *Planning for U.S. Fusion Community Participation in the ITER Program*, June 7, 2006. Available at http://www.ofes.fusion.doe.gov/News/EPAct_final_June06.pdf, last viewed July 22, 2008.

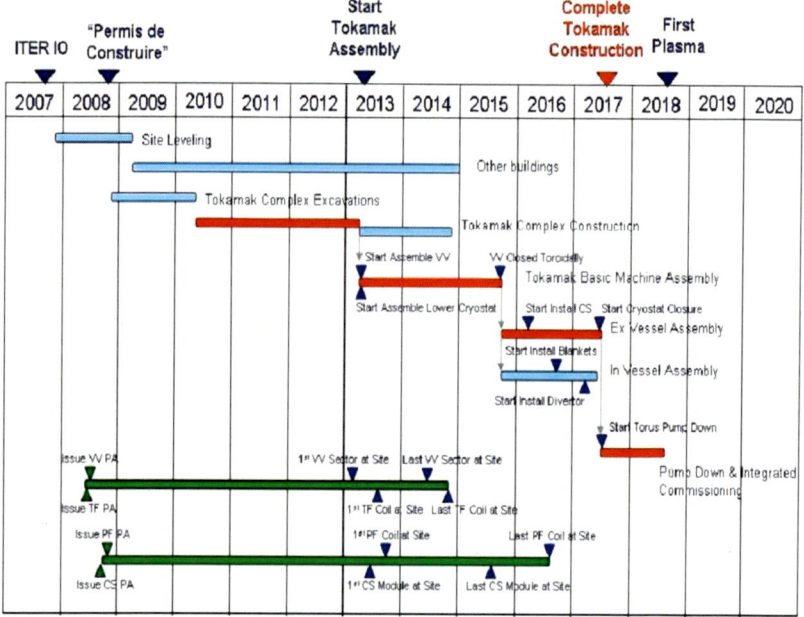

FIGURE 2.1 The revised ITER project schedule, approved by the ITER Council for planning purposes in June 2008. SOURCE: ITER Organization. Copyright by the ITER Organization. Reprinted by permission.

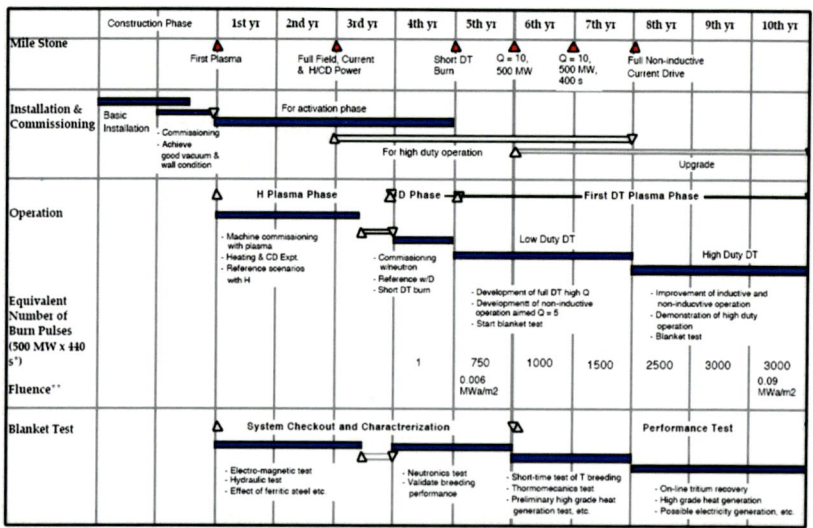

FIGURE 2.2 The current schedule of the ITER Operation Program. SOURCE: ITER Organization. Copyright by the ITER Organization. Reprinted by permission.

Key Structural Elements of U.S. Participation in ITER

The key structural elements of the U.S. participation in ITER are the U.S. ITER Project Office (USIPO), the U.S. Burning Plasma Organization (USBPO), the Virtual Laboratory for Technology (VLT), the International Tokamak Physics Activity (ITPA), and the DOE Office of Fusion Energy Sciences (OFES), as shown in Figures 2.3 and 2.4. The USIPO, the domestic project office responsible for the U.S. contributions to ITER construction, supports U.S. research and development (R&D) needed for ITER construction. The USBPO is the recently formed (2005) organization for coordinating and advocating scientific research activities in support of ITER and preparing for exploitation of ITER. OFES coordinates the activities of the USIPO, USBPO, and VLT to effectively interface with the IO. The VLT is the U.S. organization responsible for directing and coordinating engineering science and technology activities in support of ITER, including a large number of ITER R&D tasks. The director of the USBPO and the director of the VLT are the chief scientist and the chief technologist for the USIPO, respectively, ensuring close coupling of all three organizations and coupling of ITER to the U.S. scientific and engineering communities. The ITPA has been the primary international scientific coordinating body for voluntary support of ITER, identifying critical issues and facilitating joint experiments across the ITER partners. U.S. members of the ITPA are members of the USBPO, helping to ensure good communication and interaction among these groups. The ITPA, which provides a direct connection between the worldwide science communities and the IO, will soon come under the auspices of the IO. The ITPA may be viewed as the

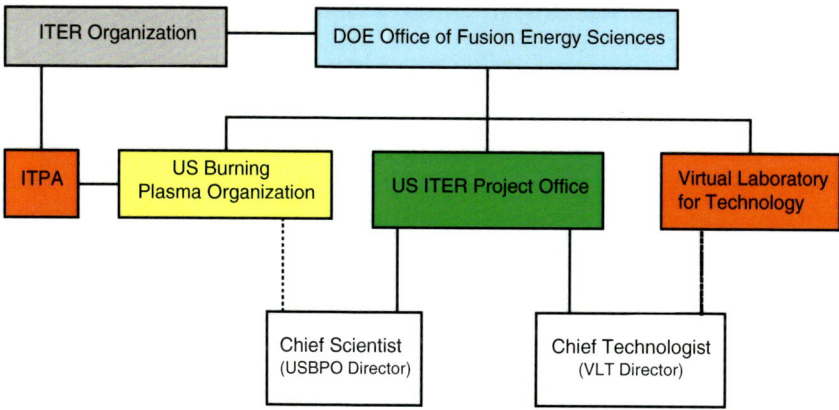

FIGURE 2.3 Major U.S. activities of the U.S. ITER effort and how they are organized. ITPA is the International Tokamak Physics Activity.

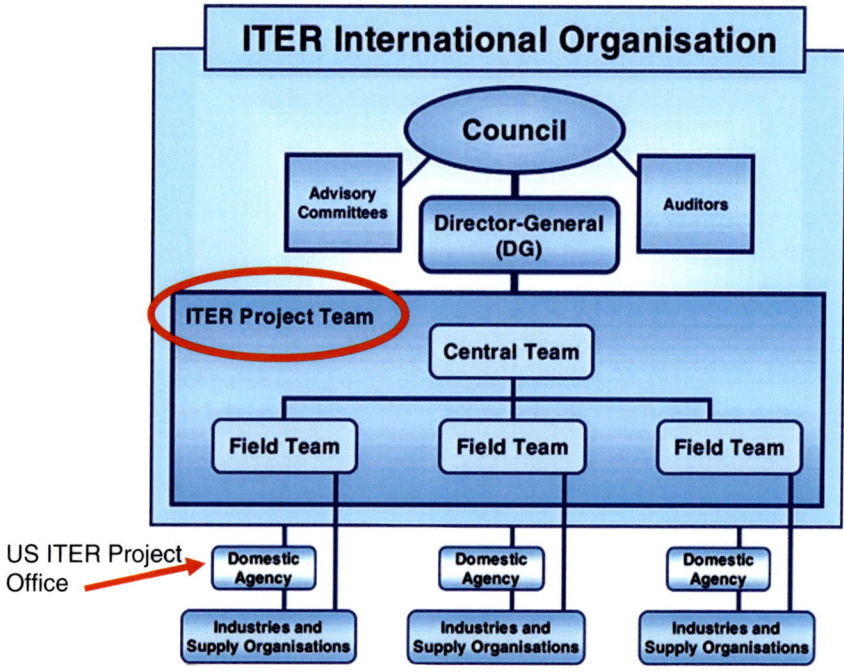

FIGURE 2.4 Overall organization of the ITER project. SOURCE: Courtesy of the ITER Organization.

precursor of the international research team for ITER exploitation. Similarly, the USBPO may be the precursor to the U.S. ITER research team or users group.

The USBPO is the key organization for participating in ITER research in the United States. It is an open organization with 289 members (as of December 2007) across the entire U.S. fusion community. The USBPO is organized into 10 research groups focused on high-priority topical areas. The group leaders meet biweekly, via videoconferencing, to coordinate, prioritize, and organize tasks on burning plasmas, focusing on ITER. The USBPO is led by a director and an assistant director, advised by a 14-member council elected from the research community. Strong leadership of the USBPO and its topical groups is key to its effectiveness. An example of this is its role in the recent international ITER design review. The USBPO topical groups identified and documented high-priority design issues, developed an objective prioritization system, and submitted the issues to the IO for consideration. The IO formed eight design review working groups, including U.S. members, to consider all the issues submitted.

Some of the issues required significant research and investigation. The USBPO, working with the members of the design review working groups, leaders of U.S. programs, the USIPO, and the OFES, identified U.S. performers for specific work packages for the review. The USBPO coordinated and completed a number of these tasks, and it prepared documentation and informative debriefings for the U.S. members of the design review working groups and the IO Management and Science and Technology Advisory Committees (both of which advise the ITER Council). Due to the effectiveness of the USBPO and other elements of the DOE plan for participating in ITER, the United States was the first ITER partner to identify performers and propose specific tasks for the United States in the design review process, ensuring that ITER would continue to be able to address the U.S. research agenda. The United States contributed 21 percent of the scientific personnel effort devoted to completing the design review tasks, even though the United States will contribute 9 percent of the construction contributions of ITER.

In addition, the IO formed an international working group to develop detailed plans for the ITER plasma commissioning and operation phases. It has established the international scientific framework and program for ITER exploitation. This includes identification of needed research developments, such as an improved comprehensive modeling capability. The USBPO is coordinating U.S. participation in this group, ensuring good communication with the U.S. research community and recognition of the U.S. research agenda. The IO plans developed by this group also provide the structure for more detailed planning of U.S. activities on ITER in the coming years.

> **Finding: The committee finds that the 2006 Department of Energy plan for U.S. participation in ITER is operating and has proven effective in beginning to coordinate U.S. research activities and the development of the ITER program.**
>
> **Finding: U.S. scientists have been well engaged in the planning for ITER, and the United States should endeavor to maintain this level of activity.**

Comparison to Analogous Efforts of Other ITER Partners

The committee believes that it is instructive to use the organizational efforts of the other ITER members as a benchmark against which to assess U.S. progress. The committee is able to comment only on the relationship of the U.S. program to the EU and Japanese research programs, which were presented in detail during its deliberations.

Overall, the U.S. international partners in ITER are explicitly organized toward developing fusion energy and a Demonstration Power Plant (DEMO). This focus gives them a clear goal for their development of fusion power. Also of note is the much larger funding profile for fusion energy research in the EU and Japan, which allows them to pursue the energy goal more aggressively. In spite of the funding differences, the present U.S. research plans for ITER are as mature as those of the other partners, and foreign partners even noted their interest in emulating the U.S. organizational structure for U.S. participation in ITER (see Figure 2.3). It is unclear at this time, however, how the elimination of funding for the U.S. first-year contributions to ITER will affect the U.S. fusion community's ability to keep its research plan abreast of the plans of its foreign colleagues.

Strong integration of the U.S. domestic research activities with the IO, through the USIPO, is facilitated by the USBPO director holding a simultaneous appointment as the U.S. ITER chief scientist within the U.S. ITER Project Office. The EU and Japanese representatives noted this arrangement as a particular strength of U.S. organization.

Finding: The committee finds that the U.S. ITER research program is at least as organizationally and technically mature as that of the other ITER participants at the time of this writing.[2]

ASSESSMENT OF THE U.S. RESEARCH AGENDA AT ITER

The research agenda at ITER that is detailed in the DOE plan addresses four overarching questions:

- How does the large size of the plasma required for a fusion power plant affect its confinement, stability, and energy dissipation properties? (large-confinement-scale physics)
- Can a self-heated fusion plasma be created, controlled, and sustained? (burning plasma state)
- Can the tokamak confinement concept be extended to the continuous, self-sustaining regime required for future power plants? (toward steady-state burning plasma)
- What materials and components are suitable for the plasma containment vessel and its surrounding structures in a fusion power plant? (fusion technology)[3]

---

[2] As of April 8, 2008.

[3] U.S. Burning Plasma Organization, *Planning for U.S. Fusion Community Participation in the ITER Program,* June 7, 2006, p. 7. Available at http://www.ofes.fusion.doe.gov/News/EPAct_final_June06.pdf, last viewed July 22, 2008.

EVALUATION OF THE CURRENT DOE PLAN

The plan details six major fusion science and technology campaigns that will be undertaken to address the four questions:

1. Integrated burning plasma science,
2. Macroscopic plasma physics,
3. Waves and energetic particles,
4. Multi-scale transport physics,
5. Plasma-boundary interfaces, and
6. Fusion engineering science.[4]

Figure 2.5 from the DOE plan presents an agenda and timeline for U.S. research and divides ITER operation into six phases:

1. Design support;
2. Pre-operations;
3. Commissioning and initial H and D operations;
4. High-gain D-T operations;
5. Modest-gain D-T, long-pulse, non-inductive operation; and
6. Fusion technology tests.

The DOE plan sufficiently explains the rationale for these research themes and how they address each research question. The plan also proposes a sequence of steps that organizes the campaigns according to the phases of ITER operation. The sequence includes the design support and pre-operations phases, which will comprise the majority of U.S. research activity in ITER over the next decade. It is important to note that to fully reap the results possible with ITER and achieve DOE's goals, the United States will have to continue to participate in ITER throughout the project's operational lifetime. The steps that the plan outlines, if achieved, would lead to fulfillment of the U.S. ITER research program objectives.

It is clear that the schedule for and approach of the U.S. research plan's science campaigns will evolve because that plan is intrinsically tied to the developing international ITER research plan, as well as to evolving domestic organizational efforts. Despite this evolution, the committee expects the four overarching research questions to remain the focus of the U.S. research agenda, given their applicability to the goals central to the ITER project itself.

A cohesive, international research plan for ITER will emerge in the future, as expected for a large international scientific project. International collaboration will be critical to the development of this research plan, and hence to the success of ITER. At the present time, it is expected that

---

[4]Ibid.

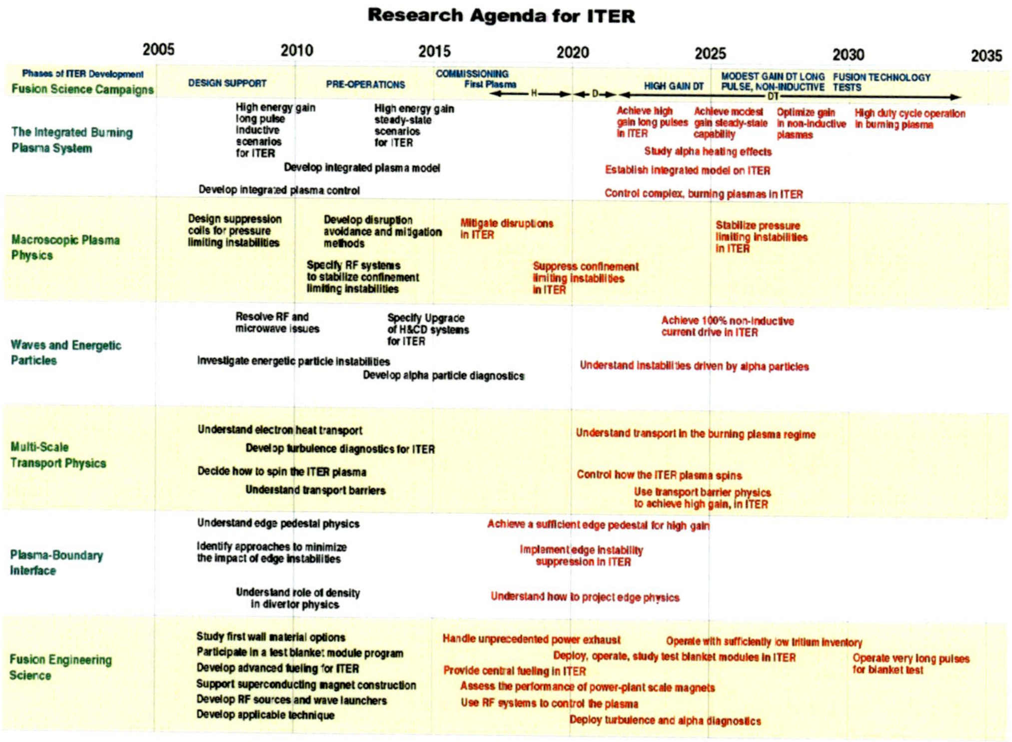

FIGURE 2.5 Anticipated U.S. ITER research agenda and timeline from the DOE plan. SOURCE: U.S. Burning Plasma Organization, *Planning for U.S. Fusion Community Participation in the ITER Program*, June 7, 2006, p. 15. Courtesy of the U.S. Burning Plasma Organization.

ITER experiments will be carried out by international teams, and so it is critical that U.S. scientists are strongly engaged in this planning process. The scientific gain reaped by the United States will depend on the ability of the nation to participate. In a nascent effort undertaken through the ITER design review, the United States has had strong participation and a significant influence. This strong participation should continue.

> Finding: The committee finds that the U.S. research program for ITER as described in the DOE plan is appropriate and justified, and the committee notes that the domestic program will evolve as the international research program is developed. U.S. involvement in developing the research program for ITER will be crucial to the realization of U.S. fusion research goals.

Alignment with DOE/OFES Goals and Previous NRC and FESAC Advice

The overarching goal of OFES is to "[a]nswer the key scientific questions and overcome enormous technical challenges to harness the power that fuels a star, realizing by the middle of this century a landmark scientific achievement by bringing 'fusion power to the grid.'"[5] ITER is a central part of the DOE/OFES program and is consistent with its stated mission of developing the knowledge base needed for an economically and environmentally attractive fusion energy source.

Earlier FESAC and National Research Council (NRC) advice strongly supported including ITER in the overall OFES program.[6,7] In particular, the committee notes the NRC *Burning Plasma* report's recommendation that "[t]he United States should participate in a burning plasma experiment."[8] ITER will address this recommendation by sustaining the hot plasma mostly through its own fusion reactions.

> Finding: The committee finds that the DOE plan for U.S. fusion community participation in ITER, in its current form, is well aligned with DOE Office of Fusion Energy Sciences goals.

---

[5]U.S. Department of Energy, *Office of Science Strategic Plan*, Washington, D.C., February 2004, p. 45.

[6]Fusion Energy Sciences Advisory Committee, *A Plan for the Development of Fusion Energy*, Washington, D.C., March 2003. Available at http://www.ofes.fusion.doe.gov/more_html/fesac/devreport.pdf, last viewed July 22, 2008.

[7]National Research Council, *Burning Plasma: Bringing a Star to Earth*, The National Academies Press, Washington, D.C., 2004, p. 4.

[8]Ibid.

Areas of Concern

The committee is concerned that the lack of funding stability will make it difficult for the United States to participate effectively in ITER, and ultimately, to have access to and thus benefit from the valuable scientific and technical knowledge to be gained from the facility. ITER is the most globally participatory science project in history, and it represents a significant step forward in the worldwide effort to develop commercially viable fusion power. But funding issues threaten to keep the United States from being a participant in this important endeavor and thus threaten the U.S. ability to capitalize on advances made at ITER. Such issues also potentially impair the U.S. ability to participate effectively in and benefit from future fusion projects that will bring commercial fusion power closer to reality. It would be a tremendous loss if the United States were unable to participate in ITER, and would severely limit the DOE/OFES ability to achieve its overarching goal.

The committee notes the wise decisions taken by DOE to keep the United States engaged, to the extent possible, in the ITER project despite budget difficulties. As the IO develops its full functionalities it will be imperative that the United States establish itself as a stable and participatory partner in order to accomplish the goals set forth by DOE, Congress, the President, and the plasma science community. The committee is concerned, however, about the ramifications that the FY2008 appropriations will have on continued progress in developing a U.S. plan for participation in the ITER project, as well as on the establishment of robust participation by U.S. scientists in the ITER research effort. As stated above, the FY2008 budget does not allocate funds to ITER as planned. Such unexpected, dramatic oscillations in commitment not only adversely affect U.S. national standing among its peers in the ITER project, but also deleteriously weaken the efficacy of careful planning that otherwise would ensure balance across the nation's broad scientific enterprise. Stable and predictable funding has been recommended in numerous NRC and FESAC reports, and this committee endorses the sapience of those recommendations.[9] Failure of the United States to meet its obligations

---

[9] See National Academy of Sciences, National Academy of Engineering, and Institute of Medicine, *Rising Above the Gathering Storm: Energizing and Employing America for a Brighter Economic Future*, The National Academies Press, Washington, D.C., 2007; National Research Council, *Plasma Science: Advancing Knowledge in the National Interest*, The National Academies Press, Washington, D.C., 2007; National Research Council, *Burning Plasma: Bringing a Star to Earth*, The National Academies Press, Washington, D.C., 2004; Fusion Energy Sciences Advisory Committee, *Review of the Strategic Plan for International Collaboration on Fusion Science and Technology Research*, Washington, D.C., January 23, 1998; and Fusion Energy Sciences Advisory Committee, *Report of the Panel on Criteria, Goals, and Metrics*, Washington, D.C., October 8, 1999.

from the outset of the ITER project will also jeopardize other countries' willingness to collaborate with the United States in future major scientific projects, possibly including a DEMO reactor. If the participation of U.S. scientists at ITER is a congressional priority, the stability of U.S. contributions to the project has to be ensured.

Finding: The committee underscores as its greatest concern the uncertain U.S. commitment to ITER at the present time. Fluctuations in the U.S. commitment to ITER will undoubtedly have a large negative impact on the ability of the U.S. fusion community to influence the developing ITER research program, to capitalize on research at ITER to help achieve U.S. fusion energy goals, to participate in obtaining important scientific results on burning plasmas from ITER, and to be an effective participant in and beneficiary of future international scientific collaborations.

Recommendation: The Department of Energy should take steps to seek greater U.S. funding stability for the international ITER project to ensure that the United States remains able to influence the developing ITER research program, to capitalize on research at ITER to help achieve U.S. fusion energy goals, to participate in obtaining important scientific results on burning plasmas from ITER, and to be an effective participant in and beneficiary of future international scientific collaborations.

Other areas of concern are noted below:

- *Gaps in the DOE plan in the planning to DEMO.* The fusion community has recently started to address issues of evolving the domestic research program. The FESAC report, *Priorities, Gaps, and Opportunities: Towards a Long-Range Strategic Plan for Magnetic Fusion Energy*,[10] reiterates requirements for a vital and forward-looking domestic research program to exploit knowledge gained in ITER through international cooperation, and it suggests initiatives to bridge knowledge gaps to DEMO. A recent NRC report, *Plasma Science: Advancing Knowledge in the National Interest*, recommended the formulation and periodic updating of a 15-year strategic plan for burning plasma research, which this committee endorses. As described in *Plasma Science*, this plan would address several issues

---

[10]Fusion Energy Sciences Advisory Committee, *Priorities, Gaps, and Opportunities: Towards a Long-Range Strategic Plan for Magnetic Fusion Energy*, Washington, D.C., 2007. Available at http://www.sc.doe.gov/ofes/FESAC/Oct-2007/FESAC_Planning_Report.pdf, last viewed July 22, 2008.

facing the U.S. magnetic fusion energy effort, in particular "the growing gap between the newer, more capable intermediate-scale facilities being built abroad and the aging U.S. facilities."[11] It will be difficult to carry out exploratory research on ITER or investigate opportunistic scenarios that may develop in the course of ITER's operational lifetime without an underpinning of smaller tokamaks within the United States and abroad. Moreover, the U.S. fusion workforce will benefit from the training that operating such devices will provide.

The strategic plan would enable the United States to maintain synergy with research coming out of ITER throughout its long operational lifetime, and thus allow the United States to contribute to and follow through on ITER research. Additionally, the DOE plan for participation in ITER will need to make clear what operational capabilities will be required of domestic facilities to support ITER if the plan for ITER is to remain synchronized with the 15-year U.S. strategic plan.

- *No discussion in the DOE plan of dissemination of ITER research activities to the broader scientific community.* Responsibility for the important role of educating the public about ITER's mission should also be made clear. The committee notes recent efforts that begin to address these issues, such as presentations at the recent meeting of the American Association for the Advancement of Science. The DOE plan will need to formulate effective strategies to establish standing lines of communication within the fusion sciences and with other disciplines, as well as with scientists and engineers in universities and industry. Although the scientific isolation of the magnetic fusion community is decreasing, much can still be done to broaden the reach of research results in the field.[12]

- *No formulation or consideration in the DOE plan of a comprehensive plan for the recruitment and training of young fusion scientists and engineers.* A related concern is the need for training of young scientists in other core disciplines, such as nuclear engineering, necessary for burning plasmas. Past NRC and FESAC studies have voiced similar concerns,[13] and DOE has taken some steps toward addressing this issue. The European Union has begun to formally implement a program to address this issue as it

---

[11]National Research Council, *Plasma Science: Advancing Knowledge in the National Interest*, The National Academies Press, Washington, D.C., 2007, p. 151.

[12]Ibid., p. 150.

[13]Ibid., p. 151; National Research Council, *Burning Plasma: Bringing a Star to Earth*, The National Academies Press, Washington, D.C., 2004, p. 7; Fusion Energy Sciences Advisory Committee, *Fusion in the Era of Burning Plasma Studies: Workforce Planning for 2004-2014*, Washington, D.C., March 29, 2004; U.S. Department of Energy, Letter from Associate Director Anne Davies to FESAC Chair Dr. Richard D. Hazeltine, October 21, 2004; and National Research Council, *An Assessment of the Department of Energy's Office of Fusion Energy Sciences Program*, The National Academies Press, Washington, D.C., 2007, p. 76.

develops its strategy to harness fusion energy.[14] The expected success of ITER in the next decade, the aging of the fusion energy workforce, and the continued workforce concerns of the U.S. and European fusion communities all dictate that consideration be given to maintaining and strengthening the U.S. workforce.

> **Recommendation: Important considerations that are not reflected in the current DOE plan for U.S. participation in ITER should be addressed during the further development of the DOE plan. These considerations include:**
> - Existing gaps in planning for a Demonstration Power Plant,
> - Dissemination of information on and the results of ITER research activities to the broader scientific community, and
> - Planning for the recruitment and training of young scientists and engineers.

ASSESSMENT OF METHODOLOGIES TO EVALUATE ITER'S CONTRIBUTION TO PROGRESS TOWARD A POWER SOURCE

Two criteria for measuring ITER's contribution to progress toward a power source have been emphasized in the DOE plan: the achievement of predictive scientific understanding, and the achievement of plasma performance characteristics for a safe, reliable, and affordable power source.

From the DOE plan:[15]

> The focus of the U.S. Fusion Energy Sciences program is the development of a predictive understanding of the fusion plasma system to support moving beyond ITER. A metric for progress in scientific understanding is whether the specific goals that collectively define the research agenda discussed above are achieved in the expected time frames. The level of agreement among theory, simulation, and experiment measures progress toward these goals. Another measure of scientific progress is the ability to use that knowledge to extend plasma performance toward that needed for fusion power. The ultimate measure of progress in scientific understanding, however, is obtained through periodic peer review of the research activities performed.

---

[14]European Atomic Energy Community, "Seventh Framework Programme of the European Atomic Energy Community (Euratom) for Nuclear Research and Training Activities (2007 to 2011)," European Union, 2006.

[15]U.S. Burning Plasma Organization, *Planning for U.S. Fusion Community Participation in the ITER Program*, June 7, 2006, p. iii. Available at http://www.ofes.fusion.doe.gov/News/EPAct_final_June06.pdf, last viewed July 22, 2008.

Plasma performance metrics are derived from specific technical goals on ITER and fusion power plant studies that have identified the major scientific and technological goals for an attractive fusion power plant. They include issues such as fusion power, fusion power gain, plasma pressure, power density, power dissipation, and neutron wall loading. Comparison of these parameters achieved in ITER to those required for a conceptual demonstration power plant provides an array of objective measures of the progress toward fusion power.

Metrics for both scientific progress and plasma performance are necessary and are mutually supportive: progress toward increasing fusion performance will likely be possible only through progress in predictive scientific understanding, and conversely, refinement of scientific understanding will emerge when predictions are compared to actual measurements on a burning plasma. The history of the fusion program shows the value of both types of metrics. Periodic peer review to measure scientific and performance progress will be important.

Finding: The committee finds that the DOE plan for U.S. participation in ITER includes well-thought-out metrics for measuring progress toward development of fusion energy as a power source.

RELATIONSHIP OF THE U.S. FUSION PROGRAM TO THE U.S. ITER RESEARCH PROGRAM

The committee considered the relationship of the domestic U.S. fusion program to the U.S. ITER research program. Considerable effort has been spent in structuring the domestic research program to be as relevant as possible to anticipated ITER operating scenarios, which serves the dual purpose of maintaining a trained workforce and maximizing U.S. ability to contribute to the planning and achievement of ITER's scientific goals. The committee underscores the importance of maintaining a vigorous domestic fusion research program.

The committee agrees with the following relevant statement from the NRC *Burning Plasma* report: "A strategically balanced U.S. fusion program should be developed that includes U.S. participation in ITER, a strong domestic fusion science and technology portfolio, an integrated theory and simulation program, and support for plasma science. As the ITER project develops, a substantial augmentation in fusion science program funding will be required in addition to the direct financial commitment to ITER construction."[16] The strong U.S. participation in the ITER

---

[16]National Research Council, *Burning Plasma: Bringing a Star to Earth*, The National Academies Press, Washington, D.C., 2004, p. 6.

design review demonstrates the importance of a vibrant base program, including personnel and facilities, that can engage in the scientific issues to be explored at ITER. It is critical that these domestic capabilities be maintained. The overall strategy of the domestic program currently is to develop a predictive understanding of the plasma science associated with magnetically confined plasmas, which the committee believed to be very appropriate to the long-term health of the U.S. fusion program, and specifically to its involvement in the ITER project. The ability to carry out detailed experimental studies of relevant plasma scenarios coupled with theory/simulation provides the framework for progress in this predictive ability, which is best accomplished with a vigorous domestic research program. Longer-term research efforts may well be directed toward reactor design, alternative approaches to magnetic confinement, and materials development in accord with DOE's strategic plan. However, each of these research areas needs to be based on improved predictive capability.

Finding: Consistent with previous National Research Council and Fusion Energy Sciences Advisory Committee reports, the committee emphasizes that a vigorous and strategically balanced domestic program is required to ensure that U.S. participation in ITER is successful and valuable for the U.S. fusion program.

3

Recommended Elements for Future Development of the DOE Plan for U.S. Plasma Science Community Participation in ITER

RECOMMENDED GOALS OF U.S. ITER PLANNING ACTIVITIES

It is clear that planning for U.S. involvement in the ITER project must be recognized as a dynamic and evolving process due to the lengthy construction phase of the experimental facilities. During the construction phase, technical advances will continue to be made, new problems are likely to be identified, and political challenges will arise at the international and national levels. Accordingly, a successful plan must display flexibility and ingenuity and must reflect continued access to a broad range of top experts from the U.S. fusion science and technology—and, more broadly, physics—community.

Consistent with previous advice, the committee suggests that the following goals be the underpinning of U.S. planning activities:

- Encouraging broad academic and industry participation in ITER, to help ensure that the knowledge gained at ITER is brought back to the wider U.S. scientific community;
- Enabling U.S. ability to contribute substantially to ITER, and maximizing U.S. ability to act on the results produced by ITER, in order to fully reap the enormous scientific and technological reward possible as a result of U.S. involvement in the project; and
- Rejuvenating the U.S. fusion workforce by the recruitment and training of young fusion scientists and engineers.

Recommendation: The committee recommends that the following goals be adopted as the foundation of Department of Energy planning activities for U.S. participation in ITER:
- Ensuring broad academic and industry participation in ITER,
- Enabling the United States to contribute substantially to and reap the rewards from ITER, and
- Recruiting and training young fusion scientists and engineers.

RECOMMENDED PROCEDURES TO FACILITATE FURTHER DEVELOPMENT OF THE DOE PLAN

The committee suggests that the following procedures be implemented to accomplish the goals recommended above:

- A long-term strategic plan for the U.S. burning plasma fusion program should be created with ITER as an important, but not the only, piece. It is essential to understand the long-term research goals in order to ensure that U.S. research activities on ITER adequately prepare the knowledge base for future fusion energy development. A broad, long-term, burning plasma fusion research strategy within the context of global fusion energy development activities will facilitate the achievement of the goals recommended above. The committee endorses the recommendation in *Plasma Science: Advancing Knowledge in the National Interest* encouraging the development of a 15-year U.S. strategic plan "for moving aggressively into the fusion burning plasma era . . . [and to] lay out the main scientific issues to be addressed and provide guidance for the evolution of the national suite of facilities and other resources needed to address these issues."[1] The creation of such a strategic plan will help the Department of Energy (DOE) ensure that the activities of the U.S. fusion program interact synergistically with the ITER project, focus U.S. research strengths, and, ultimately, bring fusion power home to the United States.

- With the maturation of planning activities, and as progress is made in constructing the experimental ITER facilities, the United States should maintain a home team to encourage broad cooperation and collaboration among all U.S. participants in the ITER project throughout ITER research and operations. The flexible and technically encompassing U.S. Burning Plasma Organization (USBPO) has been serving in this role and should continue to be relied on as an essential point of communication linking the U.S. fusion community, the International Tokamak Physics Activity (ITPA), and the DOE Office of Fusion Energy Sciences (OFES). A broadly

---

[1]National Research Council, *Plasma Science: Advancing Knowledge in the National Interest*, The National Academies Press, Washington, D.C., 2007, p. 150.

constituted home team would be most capable of bringing together elements from across the diverse U.S. plasma science community and other disciplines of physics. This home team could also help DOE and the fusion community to implement this committee's guidance.

- To maximize the value of ITER, whose technical results are to be achieved on a scale of more than 10 years, the DOE plan should consider how current U.S. plasma science facilities will support ITER research and what capabilities will be needed in the future, feeding into the long-range strategic plan for the U.S. burning plasma fusion program. Careful planning will be required to ensure the continued relevance of U.S. facilities to the ITER project and beyond.
- It will be essential for the United States to maintain a vibrant domestic fusion program, in terms of both basic research and facilities. "Transformation of the present portfolio of aging U.S. facilities into a new portfolio designed to expeditiously address key fusion scientific issues,"[2] made possible through new domestic construction or partnering in new foreign facilities, will enable the United States to maximize the scientific return on its investment and position itself to be among the world's leaders in the development of fusion power and technology. To that same end, U.S. ITER research should be guided by advice from a program advisory committee, as is research in other DOE science programs. A vibrant domestic program will also help maintain U.S. researchers' skills at the forefront of the field and stimulate interest among younger scientists and engineers and the general public.
- The current generation of large tokamaks operated by the international ITER partners plays an important role in ITER. Experiments on these devices have provided crucial input to the recent design review of ITER, and even after ITER is operational, improved scientific understanding will come from experiments done on both ITER and at other experimental facilities. The importance of maintaining and operating smaller tokamaks among the international partners is underscored by the expected cost of running ITER and its extended operational planning process. Many physics and technical issues that may arise during ITER operation can be effectively addressed on smaller devices, which will help optimize ITER operations. Unfortunately, budget restrictions in recent years have not allowed the U.S. tokamaks to operate at full capacity, limiting their contributions. These facilities are unique and represent valuable test beds for ITER research ideas. Within the scope of the entire fusion enterprise and its budget, the DOE plan should consider if it would be beneficial to increase the operating availability of these tokamaks in support of ITER.

---

[2]Ibid., p. 151.

This approach could yield a highly leveraged opportunity to improve U.S. participation in the ITER program.

Recommendation: The committee recommends the following procedures to accomplish the U.S. planning goals recommended above, and to facilitate the further development of the DOE plan:
- DOE should create a long-term strategic plan for the U.S. burning plasma fusion program within the context of global fusion energy development activities.
- The U.S. Burning Plasma Organization should continue to be an essential point of communication, and serve as a home team to encourage broad cooperation and collaboration among all U.S. participants in the ITER project.
- DOE should maintain a vibrant domestic fusion program through strong support for basic research and facilities.
- The DOE plan for U.S. participation in ITER should consider what capabilities exist and need to exist at U.S. plasma science facilities.
- The DOE plan should consider the needed operating availability of domestic tokamaks.

RECOMMENDED METRICS FOR MEASURING ROBUST U.S. PARTICIPATION IN THE ITER RESEARCH PROGRAM

Metrics Included in the DOE Plan

The committee finds that the DOE plan includes well-thought-out metrics for evaluating U.S. participation in the ITER research program. These metrics will help to inform policy makers and project leaders about the level of participation of the U.S. fusion energy program in the ITER project.

The metrics given in the DOE plan are quoted in the bulleted entries below:[3]

- "Number of U.S. researchers, students and technologists participating in ITER," and
- "Number of experiments and technology tests proposed or led by U.S. participants."

---

[3] U.S. Burning Plasma Organization, *Planning for U.S. Fusion Community Participation in the ITER Program,* June 7, 2006, p. 33. Available at http://www.ofes.fusion.doe.gov/News/EPAct_final_June06.pdf, last viewed July 22, 2008.

The level of participation of U.S. researchers in the ITER project, U.S. contributions to ITER experiments, and related research are indicative of the vitality of U.S. involvement in the ITER research program. Conversely, these metrics will also provide insight into the contribution of ITER research to the U.S. fusion energy research program. As U.S. researchers continue to participate in ITER research and development activities, they will bring back the knowledge gained and apply it for future advances in the U.S. base program.

- "Achievement of scientific and technology milestones on ITER."

ITER is a scientifically, technologically, and organizationally challenging project. Setting and then meeting ambitious, yet realistic, milestones will not only demonstrate progress toward achieving the planned research but it will also support and encourage the international partners in ITER.

- "Number of research and technology publications on ITER produced by U.S. participants," and
- "Citations of U.S. publications."

Bibliometrics is a widely recognized method of evaluating research impact, and it will help policy makers and researchers to assess the health of U.S. participation in the ITER project and research. The committee emphasizes that the citation of U.S.-based research appearing in publications from ITER is also a valuable metric because it directly reflects the U.S. influence on ITER research. However, program managers should not rely on bibliographical figures alone, but should run complementary analyses using the metrics outlined in this section and the next. It is understood that data on publications are influenced by a variety of factors and can vary from project to project, and so having a suite of assessment tools is critical.

In fact, the U.S. fusion community is already robustly engaged in the ITER research program and in the design and construction process at all levels, through the USBPO, the Virtual Laboratory for Technology (VLT), and the U.S. ITER Project Office (USIPO) and their close affiliation with the other ITER organizations. Recently, the U.S. program has participated strongly in the international ITER design review, organized by the ITER Organization, to complete the ITER baseline design. The USBPO has applied some of the metrics in the DOE plan to recent activities, with the following results:

- 278 U.S. researchers from 49 institutions are members of the USBPO. Approximately 124 U.S. researchers participated directly in the ITER design review.
- Approximately 50 percent of the experiments planned for 2008 on the largest U.S. experiments (C-Mod, DIII-D, NSTX), taken together, are in support of ITER.
- U.S. scientists constitute 73 of the 273 authors (27 percent) of the nine articles documenting *Progress in the ITER Physics Basis*, published in the journal *Nuclear Fusion* in 2007.[4]
- U.S. scientists constitute 30 of the 68 authors (44 percent) of the 13 articles on diagnostics for ITER and burning plasmas, published in a special issue of the journal *Fusion Science and Technology* in 2008.[5]
- U.S. scientists were the lead authors on 10 of the 65 papers on ITER at the 2006 IAEA Fusion Energy Conference (Chengdu, China) and were co-authors on an additional 9 papers.

These activities came about from proactive engagement by the USBPO, the USIPO, and OFES, and provide early evidence that the DOE plan is working well.

When evaluating the results of applying these metrics, it is important to compare the results to the 1/11th share of the project that the United States has agreed to contribute. It will be equally important to bear this in mind for future evaluations. The USBPO's evaluation provides early evidence that the United States has been engaging effectively in international ITER planning activities, although this level of participation will remain contingent on U.S. support for the project.

Finding: The committee finds that the DOE plan includes well-thought-out metrics to measure the robustness of U.S. participation in the ITER program.

Recommended Additional Metrics

The committee recommends that five additional metrics be considered during the future development of the DOE plan for U.S. participation in ITER, namely:

---

[4]International Atomic Energy Agency, *Progress in the ITER Physics Basis*, Vol. 47, No. 6 of *Nuclear Fusion*, June 2007, IOP Publishing, Vienna, Austria.

[5]American Nuclear Society, *Fusion Science and Technology*, No. 2, February 2008.

- Periodic evaluation by expert and knowledgeable members of the scientific, engineering, and industrial community regarding the U.S. return on its ITER investment.
- Periodic assessments by independent, external bodies of the effectiveness of domestic project management.

The committee stresses that peer review evaluations of U.S. participation in the ITER project could provide the most reliable measure of robustness. Until the ITER Organization is fully staffed and the international research plan is set in motion on an operational ITER, numerical metrics may not be sufficient to judge the robustness of U.S. participation and the ITER project's effect on the domestic U.S. fusion program. Similarly, to properly gauge organizational progress in establishing an effective participatory relationship with the ITER Organization's management structure and the project that it runs, independent advisory assessments will be needed. These assessments will give U.S. decision makers early and independent insight into the vitality of U.S. involvement.

- Balance in the fraction of U.S. published research conducted on ITER according to authors' institutional affiliations (university, national laboratory, and industry).

The DOE plan's metrics are concentrated on measuring the level of activity of the U.S. program but do not characterize that activity. Strong U.S. participation in ITER will require the involvement and coordination of researchers from universities, national laboratories, and industry. Ensuring that a healthy balance is struck will be critical. This balance will need to be determined by an advisory committee.

- Number of research and technology publications documenting results obtained on ITER that are cited by or produced in collaboration with U.S. researchers, students, and technologists across U.S. plasma science and physics.

Although the metrics currently included in the DOE plan indicate the level of involvement of U.S. researchers in the ITER project itself, they do not provide insight into the unique synergistic effect that ITER research "coming home" will have on the U.S. base program. The fusion community and DOE expect that research conducted for ITER will provide a tremendous intellectual boost to the U.S. base program, and having a tool to measure this invigoration will be valuable to policy makers.

- Achievement of predictive capability that will offer another effective measure of the success of the U.S. ITER program.

If the United States demonstrates the capability to predict ITER operating parameters, as well as other important measures such as component lifetime or suitability in a commercial fusion device, it will be a good indication that the United States is participating robustly in the ITER research program, although expert panels must necessarily address such questions in a peer review process because of their technical complexity.

Recommendation: The committee recommends that the following five metrics be considered for inclusion during the future development of the DOE plan for U.S. fusion community participation in ITER:
- Periodic evaluation by expert and knowledgeable members of the scientific, engineering, and industrial community regarding the U.S. return on its ITER investment.
- Periodic assessments by independent, external bodies of the effectiveness of domestic project management.
- Balance in the fraction of U.S. published research conducted on ITER according to authors' institutional affiliations (university, national laboratory, and industry).
- Number of research and technology publications documenting results obtained on ITER that are cited by or produced in collaboration with U.S. researchers, students, and technologists across U.S. plasma science and physics.
- Achievement of predictive capability, to be evaluated by peer review.

Appendixes

Appendix A

Letter of Request from the U.S. Department of Energy

Under Secretary for Science
Washington, DC 20585

August 10, 2006

Dr. Ralph Cicerone
President
National Academy of Sciences
Mail Stop 822
500 Fifth Street
Washington, D.C. 20001

Dear Dr. Cicerone:

The Energy Policy Act of 2005 (EPAct), Section 972, authorizes U.S. participation in ITER, and directs the Department, in consultation with the Fusion Energy Sciences Advisory Committee (FESAC), to develop a plan for the participation of U.S. scientists in ITER. Additionally, Section 972 (c)(4)(B) of EPAct specifically directs the Department of Energy to request a review of the plan by the National Academy of Sciences.

On behalf of the Secretary of the Energy, I have enclosed a plan titled "Planning for U.S. Community Participation in the ITER Program" prepared by the U.S. Burning Plasma Organization under the guidance of the Office of Fusion Energy Sciences. This plan was discussed by the FESAC at its meeting on June 1, 2006. The FESAC review and endorsement of the plan is enclosed.

If you have any questions, please contact me directly at (202) 586-0505 or Dr. James F. Decker at (202) 586-5434.

Sincerely,

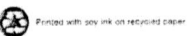

Raymond L. Orbach
Under Secretary for Science

Enclosures:
Plan
FESAC Comments

cc:
D. Shapero, NAS

Appendix B

Meeting Agenda

FRIDAY, DECEMBER 14, 2007, WASHINGTON, D.C.

Closed Session

7:30 am	Breakfast available
8:00 am	Committee discussion
10:00 am	Break

Open Session

10:15 am	DOE perspectives and plans for engagement of ITER E. Oktay, DOE/OFES
10:50 am	Discussion Committee and Guests
11:00 am	Perspectives from the Office of Science and Technology Policy (OSTP) K. Beers, OSTP
11:25 am	Discussion Committee and Guests
11:30 am	ITER Organization engagement of member states and research agenda D. Campbell, ITER Organization, Fusion Science and Technology Department [by telephone]
12:20 pm	Discussion Committee and Guests

12:30 pm Working lunch
1:30 pm   EU participation in ITER and research plans
          J. Pamela, European Fusion Development Agreement
2:05 pm   Discussion
          Committee and Guests
2:15 pm   Japanese participation in ITER and research plans
          S. Matsuda, Japan Atomic Energy Agency
2:50 pm   Discussion
          Committee and Guests
3:00 pm   Break
3:15 pm   Plans for U.S. engagement with ITER Organization
          N. Sauthoff, Oak Ridge National Laboratory (ORNL)
3:50 pm   Discussion
          Committee and Guests
4:00 pm   Engagement of U.S. plasma science community in ITER research and reflections on USBPO plan
          J. Van Dam, USBPO
4:35 pm   Discussion
          Committee and Guests
4:45 pm   Activities of the USBPO Panel on Long-Range Burning Plasma Program Planning
          E. Marmar, Massachusetts Institute of Technology
4:55 pm   Discussion
          Committee and Guests
5:00 pm   U.S. engagement in ITER technology research
          S. Milora, ORNL
5:35 pm   Discussion
          Committee and Guests
5:45 pm   General discussion
          Committee and Guests
6:00 pm   Break and depart for dinner
          Committee and Speakers
6:30 pm   Working dinner
8:00 pm   Adjourn

SATURDAY, DECEMBER 15, 2007, WASHINGTON, D.C.

Closed Session

7:30 am	Breakfast available
8:00 am	Committee discussion
10:00 am	Break
10:15 am	Continued discussion

11:30 am	Working lunch
12:30 pm	Continued discussion
2:30 pm	Break
2:45 pm	Continued discussion
4:30 pm	Adjourn full meeting